Welcome to the Anthropocene

Alice Major Welcome to the Anthropocene

 The University of Alberta Press

Published by

The University of Alberta Press
Ring House 2
Edmonton, Alberta, Canada   T6G 2E1
www.uap.ualberta.ca

LIBRARY AND ARCHIVES CANADA
CATALOGUING IN PUBLICATION

Major, Alice, author
    Welcome to the anthropocene /
Alice Major.

(Robert Kroetsch Series)
Poems.
Issued in print and electronic formats.
ISBN 978-1-77212-368-5 (softcover).–
ISBN 978-1-77212-397-5 (EPUB).–
ISBN 978-1-77212-398-2 (Kindle).–
ISBN 978-1-77212-396-8 (PDF)

    I. Title. II. Series: Robert Kroetsch
series

PS8576.A515W45 2018   C811'.54
C2017-906797-4
C2017-906798-2

First edition, first printing, 2018.
First printed and bound in Canada by
Houghton Boston Printers, Saskatoon,
Saskatchewan.
Copyediting and proofreading by
Peter Midgley.

A volume in the Robert Kroetsch Series.

The University of Alberta Press
gratefully acknowledges the support
received for its publishing program
from the Government of Canada, the
Canada Council for the Arts, and the
Government of Alberta through the
Alberta Media Fund

For David,
again and always

Contents

Prologue

In medias res

Alas poor child, you're born
in medias res — the stage is set
with swirling depictions of a globe
in panic, small rainbow-coloured frogs
hopping into oblivion,
a scene of smoggy atmospheres,
vast gyres of plastic churning
in the ocean, Scylla and Charybdis,
sailors screaming from their boats,
soldiers raising fists, battle-dressed
for costumed wars.

And you have got to figure out the script.
It's that recurrent nightmare
of being unprepared, of never
having studied and now it's curtain time.
That dream is just our human
situation — the only plot we've got
in this play without an author.
We're writing it ourselves.

And I can't help you. I am just
another figure in the chorus
of greying heads, wringing her hands
or pointing to a star.
Sorry to be useless, but that
is what we are.

Dear child of fortune, born today
into the middle of things,
break a leg. Don't look for gods
descending in a basket,
or prompters in the wings.
Declaim one memorable soliloquy.
Turn a spotlight. Or pick up
pelting litter from the stage.
There is no ending, happy
or otherwise. Just play your part.

Welcome to the Anthropocene

Welcome to the Anthropocene

1.

> *In pride, in reas'ning pride, our error lies;*
> *All quit their spheres and rush into the skies.*
> — ALEXANDER POPE, '*An Essay on Man*'

Welcome, you line of dogs whose sizes span
the gamut from gargantuan Great Dane
to Lilliputian chihuahua. And welcome,
Freckles the goat, to our expanding album
of post-natural creation — your milk
seeded with proteins from arachnid silk,
orb-weaver spiders woven in your genes.

Welcome to the Anthropocene,
Black-6 Mouse, your myriad descendants
scrabbling in labs, bred for their resemblance
to each other and to us: a tendency
to age-related hearing loss; efficiency
in breeding but erratic parenting;
willingness to drink booze; inheriting
a sensitivity to pain and prone
to biting. Your murine chromosomes
are known to the smallest nucleotide
and carbon-copied endlessly — a kind
of immortality that we've bestowed
by making you so uniform you flow
indistinguishable as plastic beads.

Welcome transgenic zebrafish. Your shades
of trademarked colours — *Starburst Red*,
Electric Green, and *Cosmic Purple* — bred
to decorate café aquariums
in colour schemes to match our rainbow whims.

Welcome, also, to the dumbed-down denizens
of Bottle 38 — you specimens

of *Drosophila* we have knocked about
with mutagens designed to cancel out
the gene-made protein that would let you learn.

We might ask what it is that *we* have learned
from setting out this biological array,
our manufactured freak show. Outré
artificial creatures, genetic lines
we've crossed and re-crossed far too many times
in our compulsive drive to flout
all natural order. The dogs with snouts
so short that they can hardly breathe.
Hens bred to dreadful uniformities
of shape and size, no variations wanted.
We itch to think of spider bits implanted
in lactating mammals. This mad caprice
of test tubes makes us shudder. *Poor beasts.*

Poor beasts. We have been rattling The Great Chain
of Being. Feckless godlings, we're inflamed
by our capacities, creating mice
in our own image, trying to entice
genes to jump with tickled transposons
and scraps of virus. We have good reasons,
so we claim, for meddling: improved tomatoes,
new vaccines or fuel for our autos.

    But is it not the sin of pride
that we express? Hubris personified?
We will not admit to limits, will not
hold back rash action for sober thought
about the unintended consequences
of what we've set in motion, the immenseness
of what our broken chain links might turn loose —
chaotic spiralling of feedback loops,
exponential impacts.
                    We don't believe,
these days, that God rebukes presumption. We've
given up on that great, heavenly anchor

on which the chain depends, don't fear the anger
of omniscience. But we are not gods
who know the outcomes that we set abroad.
And, though hubris always teeters at the very edge
of tragedy, we will not look up from the page
that we are scribbling too busily
to think of ends. Meanwhile, our bestiary
of created things looks sadly back at us
from beyond the paling of our apparatus.

2.

   All are but parts of one stupendous whole

And yet our pug dogs snore contentedly
beside our beds. They're temperamentally
incapable of resenting two square meals
a day, a waddle in the park. And Freckles feels
quite normal, thank you. Her silk-kids frolic
and munch their premium hay. Alcoholic
mice are rarely on the street panhandling crumbs.
The GloFish in their safe aquariums
don't care that they're conspicuous.
No predators to dodge. And as for us,
our chromosomes are stuffed with our mélange
of post-natural genetics. It's just as strange,
really, that our box of HOX and PAX
is shared with chimp and fruit fly, all these swaps
transferred from philandering bacteria:
the opsin molecules that line our retina,
the collagen granted us by sponges,
the histone strands we get from fungus
and share with lamprey, shark and kangaroo.

The Great Chain of Being links us — true.
But it's not a ladder to the angels.
It's a horizontal loop that rearranges
life repeatedly. It's still ongoing —

protists leaving bits behind, a snowing
of genetic flakiness that's *not* man-made.

Not chain, not ladder — we're relatives, a clade.
The planet's been a boiling Petri dish
of life since long before those lobe-finned fish
crawled shorewards half a billion years ago
to breathe in air, dragging a portmanteau
already stuffed with DNA in use
by clams and algae and handed on to us
to rummage through. Why get excited
over corn, just because we've hyped it
with genetic transfers from bacteria?
Corn genes are *founded* on bacteria!
There is no species we can label 'pure'
and fence around eugenically, secure
against pollution. Not even humans,
latest offspring of the protozoans.

Still, it behoves us to take care, to choose
wisely. We could love a pug whose nose
was long enough to breath with, raise fowls
for values other than to be our victuals.
Though it's excessive to excoriate
all geneticists, it's appropriate
for us to demonstrate humility.
We are not angels. But we're family.

3.

> Then in the scale of reas'ning life, 'tis plain
> There must be somewhere, such a rank as man.

Life bubbles into every crevice, niche,
and hairline crack — a ramifying mesh.
Continents drift off, islands rise from sea,
reefs build on reefs. Forest canopies
and cavern mazes, tundra, jungle, shore —

every landscape will invent (from chance spores
and accidental founders) the creatures
that it needs to fill its lap. The grass eaters
and flesh shredders, the vast stampeding herds
and the teeth that hunt them down, the birds
that interweave their voices through the chorus
of dawn. And, always, creepers through the forest
undergrowth — whether the stirring branches
overhead are cedar or seaweed, pampas
or algae, taiga or the cells that line
our gut. In every biome, creatures glean
and browse, hunt and hide and specialize —
partners in the great translation enterprise
from chemistry to useful energies
for living. And somehow, all this leads
to an open space, a function we could fill
because the gap is there — a gap we call

intelligence. Not a separate limb
or magic faculty inserted in
our brains. Rather, an elaboration
(through millennia of tiny, patient
trials-and-errors) of the skills required
by any animal that has been wired
for movement in the world. A widening
of memory, from simply recognizing
patterns here and now to narratives
of lived experience. In turn, that gives
an opening to calculate the future —
to plan, imagine; to be astuter
than the other beings at the mercy
of storm and season. Nothing unearthly
decrees our brains' expansion into tools
and storytelling, just the constant pull
of that open door before us. Nor did Nature
need great leaps to build a social creature
that's able to do more in concert than
a single memory or lifespan can —

a creature that could build through generations
(layer on layer, scrap by scrap, calculation
after calculation) an edifice
of record, shared, preserved and pressed
into bedrock. Learning that survives,
compacted, the way the tiny lives
of foramens rain down on ocean beds
and settle into limestone, or the myriad
coralliferous remains of animals
raise giant reefs from minute burials.

12   Perhaps it could have been the clever corvids
who got here first, heading up the scorecard
of cognition, using their nimble beaks
to master tools, learning new techniques
for modifying their environment,
working the muscle of intelligent
cooperation. The ravens, who already call
in croaking protolanguage, could evolve
the broader pattern of symbolic speech.

              Or perhaps our niche
might have been filled by the invertebrates
(who started long before us), and the gate
pushed open by a suckered tentacle,
a smarter cephalopod. Chemical
riffs and rattles, changes, might have loosed
cascading adaptation and put to other use
the scintillation of chromatophores.
Imagine colours used for something more
than flares of anger, urgent camouflage.
Imagine a vivid, silent language
sweeping over skin, instinct's dictation
translated into willed communication.
And then an ocean floor built up with cities,
herded fish-flocks, the patternicity
of gardens, turrets, standing stones, machines —
all jointly engineered. It might have been.

But primates reached it first. That opening.
Scarcely knowing what was happening,
not planning domination, nor with any goal
beyond survival. We were like a coal
in a nest of moss, ready to ignite
and carried to another campsite.

4.

From Nature's chain, whatever link you strike,
Tenth or ten thousandth, breaks the chain alike

Welcome, welcome to the Anthropocene
raccoon, coyote, house mouse, peregrine,
squirrel, red fox, *Rattus norvegicus* —
all you creatures who can live with us,
being sufficiently plastic to adapt
and thrive upon our handouts, urban crap,
suburban rubbish dumps and garbage cans.
Welcome Canada goose, taking your stand
(all five million of you) on our parks
and golf courses — you avian oligarchs
hissing at our dogs, dropping grey-green turds
on swathes of grass. You're what we've deserved
after we've homogenized the landscape
planet-wide. Our broad foot eradicates
the little islands of ecology,
the disappearing rare, the melody
of the threatened: red-eyed vireos
piping plovers, grasshopper sparrows —
all the small, sweet, uncompetitive.

Immured in cities, we forget we live
on a planet that is more inventive
than ourselves. Her secrets are undreamt-of,
even now — her hidden leaves and worms,
her microbes, her amphibians. Yet we churn
her soils, her ocean depths, her streams

like the thwacking paddles of a dough machine.
Worldwide, our cities rise as uniform
as mass-produced white bread. We transform
the richly variegated niches
into starved soil for weedy species
like ourselves. Mown, shorn vegetation.

Chronically impoverished, yet unchastened
we think the gadgetry we've gained redeems
our losses. Why should we miss one small, green,
leaf-shaped frog, gone from a distant tropic
half a world away? We are too myopic
to see this slender loss might mean a space
is closed, a possibility effaced.

5.
Atoms or systems into ruins hurl'd,
And now a bubble burst, and now a world.

To all you entries in the global database
of life: welcome. Welcome to this hyperspace
during which humanity has hacked
into the planet's history. In this tract
of ad hoc coding, we're running trials
like half-assed systems analysts whose files
have never been backed up, reckless geeks
who don't know when we've pressed 'delete'
once too often.

                        Still, we might be content
on a planet with no great auks or elephants,
polar bears or pandas. How often do we meet
Sumatran tigers on our city streets
(or want to)? We could simply look
at legendary beasts in picture books
or videos. They're nice-to-haves, not musts
for daily life. As for rhinoceros,

white shark or Orinoco crocodile,
who'd care for living with one, cheek by jowl?

We don't mourn the passing of the mammoth
every morning, nor the vanished giant sloth,
even if our weaponry inventions helped
to push them off extinction's sharp-edged shelf.
In fact, we've benefitted from the cull
of evolution. We'd not be here at all
if dinosaurs had not turned up clawed toes
and left. Yes, it's too bad about the dodos,
but there are many other lineages
of pigeon. The earth still manages
to maintain its total biomass. That bulk
may shift from balanced muscle to a pulp
of sagging flab around the waist; it matters
not the least. There are as many creatures
living on the planet as have ever been
— even if a lot them are hens.

But fear is growing in us (like a gas
after too rich a meal) that we have passed
some threshold — that we may be rendering
earth derelict, a disaster ending
not just giant pandas but ourselves.
A fear we're blocking earth's escape valves
and bio-sinks. Many will dismiss the question —
they say it's just a touch of indigestion,
we'll be fine. Besides, they say, it isn't us —
one good fart of forest-fire exhaust
dwarfs all the output of our vehicles.
Still, doubt's sour odour lingers in our nostrils
like effluvia wafting from our garbage dunes.
Our conurbations spread their plumes
of carbon far beyond the city limits,
and our roaring engineering mimics
volcanic-level belches every day.
Sober citizens consider ways

to plan for rising tides and surging storms
as polar ice caps melt and our world warms.
We design deployable walls, but feel
as if we were the child in some old tale
of dikes and imminent disaster,
sensing that the cracks are spreading faster
than adults (waking finally) can mend
with chips of silicon and bags of sand.

Life has done this before — tipped the balance
when microbes poured a poisonous chalice
of oxygen into their air. Dumb actors
in their own demise, *Cyanobacter*
out of control, replicating round the globe.
Their own waste product choked those anaerobes
until they couldn't take it any more
and died — or else cling on in corridors,
lining guts and hydrothermal vents,
leaving the open seas and continents
to other life forms.

Many folk dismiss
this history, insisting *We can fix*
anything, we're smarter than bacteria.
There isn't any reason for hysteria.
We'll plant some trees. But do we really want
to take the risk? We don't seem intelligent
enough to work together, work through
our rifts and schisms. More likely we will do
little more than flap our techno-wings.

Will it be our place in the scheme of things
— with all the virtual flim-flam we've installed —
to burst the blown-glass bubble of our world?

6.

from the green myriads in the peopled grass

Our vaulting crania, our vaunted brains —
these advantages, we feel, explain
our value, status, function. Thus, we stand
above a mindless landscape, in command.
From our cloud-capped towers, consciousness
looks out through window slits, past buttresses.
We know, and know we know. This is the power
of selfness. It sustains the lonely tower
in which we analyze and plan. We know
the past, and thus are impressarios
of future — its flags fly from the parapets
on which our terra-forming intellect
strolls out, surveys horizons, catalogues.

And in our image, we've imagined gods —
taller towers of knowledge, seeing farther
than ourselves, a conscious judgment larger
than our own.

                    But now consider slime mould.
Ancient eukaryotes, perhaps as old
as any form of life on land. They drift,
amoeboids, singletons, through soil, then shift
the gears of being when the times get tight.
They signal aggregation, pull close, unite
into a purposeful slug, intent
on reaching better sites. And then turn plant —
re-morph, erect a stem. Some cells sacrifice
their hope of future this way, pay that price
for collective-style survival of the group.
The others cluster at the tower's top,
assembling through a chemotactic arc
to form a globe — a puffball sporocarp
from which a light diaspora of dust
starts life again, a drifting exodus.

Is this perhaps our role? To climb
the tower of consciousness and then become
a scatter of gametes, a kind of seed,
DNA the universe may need

— or not. So much order is created
without brains or their associated
nattering narrators. Nature solves
her vast equations without fuss — the scrawls
of protein folding, evolving puzzles
posed by careering quantum particles.
The universe can calculate our weight
precisely — does not need to estimate
or fret about the meter's accuracy —
and measures up the force of gravity
(which ties us to the farthest entities
of space) unfazed by its immensity.
Meanwhile, our plodding, conscious calculations
can barely cope with fluctuations
stirred into the paths of circling planets
by the smallest, palest, lunar fragments.

Our conscious brains could never meet the needs
of the green grass and its peopled myriads.
They're fine without us, creating structure,
nested relationships, complex sculptures
of wasp and termite, the soil societies
of mycelium and root, the varieties
of symbiosis that marry leaf and worm.

Could such *unthinking* be a god's true form?

7.
    But ALL subsists by elemental strife

Welcome, now, *Homo sapiens sapiens*,
to this your era, nailed to all the stations

of your crossroads. Welcome ideologues
of every stripe, thrusting your catalogues
of sacred text against the camera lens.
Your fundamentalist array extends
throughout this large museum gallery
of post-natural pyroballogy:
a diorama, where jihadis pose
with hoisted rifles, Catholic Provos
plant a car bomb, Orange Loyalists
raise pistolled paramilitary fists.
Boko Haram warriors, their faces swathed
in khaki camouflage decreed by God
crouch next to costumed replicas
of neo-Nazis tattooed with swastikas.

Along this wall, all the ultra-orthodox
are grimly cased in glass with pens and inkpots:
Haredim with holy scrolls, righteous Baptists
rifling through bibles, austere Wahabists
— all studying, studying intently
their prophets' words. All convinced of entry
into heaven by entries codified
long centuries ago, decrees inscribed
on granite pages. Even when such scholars
are slapped upside the head by waves of knowledge
sloshing in, they keep their certainties
wrapped tight in waterproof hyperboles
and held above the tides of information
that could drown them in its channels. They shun
the thunder militant of atheists
(as certain as themselves) and they resist
all gentler doubts, claim *Scripture is inerrant!*
(Ours is anyway. The others aren't.)

Round this rotunda, all the generals
are on display. They're also evangelical
about the in-group markings of their tribes:
insignia and battle ensigns, stripes

and stars, battalion bars and badges, braid
— *imperial blue, old glory red* —
dangling with medals. They're flanked by flags
of national identity, the tags
of partisan allegiance. All united
by severance, togetherness incited
by our complicated need for hate.

Meanwhile, welcome to the iron gate
and worn-down steps of this repository
all you refugees, from territories
wrecked by havoc and the power of wars.
And your descendants, washed up on the shores
where they will cling with pride of ownership,
denying entry to their landing strip
to all those bobbing in the seas behind.
We got here first. No room. We're doing fine.

And welcome to the billions muddling through,
hoping for the best, not sure what to do
to make it happen, caught up in our lives,
using brains that helped us to survive
millennia ago. These toolkits crammed
with handy hammers, wrenches, second-hand
contraptions for extracting data from
the rush of wavelengths roaring on
around us: rules of thumb and biases,
buried assumptions no one realizes
we're reaching for when trying to assess
the stranger at the door, the speech, the dress.
What template makes this unknown shape make sense?
Do I dare trust that face? Expedience
take precedence above exactitude
and all our pictures of the world are crude.

Unaware, we live our lives according
to scripture written in our genes. Recording
angels of our DNA inscribe

commandments for belonging to a tribe.
We are descended from the groups most able
to cohere and trounce the others; to cradle,
care, communicate among themselves
and then go off to visit merry hell
on any other troop that tries to claim
their territory. All this with a brain
that doesn't realize its gaps and patches —
the leaps, elided details, makeshift matches
inherent in the maps it sketches. This
stubborn trust in our analysis
comes from an underlying default plan
we never notice, the wiring diagram
we're stuck with.

            The craving to be 'in'
is laid down in that mesh. This is the sin
we can't escape, the angel's two-edged blade,
the human nature that we can't evade.
We need each other. Deep within our bones
we know that none of us survives alone.
But oh, the constant damage that it does,
our ineradicable love of clubs.

8.

Hope springs eternal in the human breast

So where does all this leave us? Where do we
belong? Essential to the potpourri
of universe? Or just peripheral?
A whiff of carbon on a rolling ball
around an ordinary star? We're not
the centre of the cosmos, as we thought.
In the middle, yes — halfway up (or down)
the logarithmic scale of size that runs
from godly, intergalactic vastness
to Planck-length prick. But the web that grasps us

doesn't need us. So what is our purpose?
Are we just fruit flies batting at the surface
of a lonely bottle? Is it the doom
of all our effort simply to become
the lonely keepers of an isolated
termite mount in orbit, a co-created
system of ventilation and control,
our little lives assigned to castes and roles.
A mildly interesting experiment
in eusociality, but irrelevant
to the giant fields of force that drive
galaxies, black holes and quasars to survive.

We could even blow our test tube up —
It wouldn't matter, not a monkey's nut.
The cosmic roil makes bangs far bigger than
any feeble planetary species can;
the average galaxy from which we're spawned
would hardly even notice we were gone.
Nor are we needed here — should be deposed
from this small dominion. For, heaven knows,
we're annoying monarchs for the planet,
stripping its riches. If we weren't on it,
creating our crust of crud so crazily,
the biosphere could breathe more easily.

Yet our hope springs eternal — surely we
matter more than matter. Theology
informs us that we have immortal souls
somehow. Somewhere. (Perhaps as vacuoles
that store divinity inside our cells?
As aureoles we drape around ourselves?
A scent of roses that will pass from us
with our last breath? A stamp, mysterious
yet unmistakeable, that marks
a random clutch of atoms with the spark
of singular identity?) Naïve
or scholarly, most of us believe

we've got these slivers of forever, though
we're fuzzy over how they work and know
the devil's in those details. Souls don't make
much sense, we fear, but still we can't forsake
the comfort of this butterfly confection —
the hope that hatches from our (mis?)conception.

9.

Lives through all life, extends through all extents,
Spreads undivided, operates unspent.

Now, welcome to the Anthropocene
you battered, tilting globe. Still you gleam,
a blue pearl on the necklace of the planets.

This home. Clouds, oceans, life forms span it
from pole to pole, within a peel of air
as thin as lace lapped round an apple. Fair
and fragile bounded sphere, yet strangely tough —
this world that life could never love enough.
And yet its loving-care has been entrusted
to a feckless species, more invested
in the partial, while the total goes unnoticed.
Our inconvenient hearts, so focused
on what is near — the pet dog's suffering
but not the world's. Our attention sputtering
like fading flashlights. Meanwhile leaders wave
light sabres wildly, as they try to carve
our common interests into fragments.
Our coherent will to act turns stagnant,
our batteries go flat. We cannot see
beyond the dark in our vicinity.

Dear planet, we might find illumination
in you, and lessons in a murmuration
of starlings, shape-shifting veil of wings
in evening air. Its flow of change begins,

not with a leader striking out a path
towards a goal, dragging along a swath
of followers, but from a turning in;
guidance not from centre, but from rim.
A few birds at the edge respond to danger,
plunge of predator, and the remainder
reorient to synchronize their flight
as quickly as the particles of magnetite
shift in a magnetic field. As fast as
thought goes rippling through synapses
that don't quite touch.

            Then think how everything
does touch. Our universe comes blossoming
out of a vacuum that is not void
but plenum, boiling substrate, *being*, buoyed
by its own unceasing, fizzing spin and spit.
Tension, potential, a tinkerer's kit
of fields and forces, virtual bosons joined
into a charged dimension where every point
is a world defined by multiple descriptors:
gluon field and gravity, the raptures
of light and magnetism intertwined,
attraction, repulsion balanced and combined.
And from the layered, interlocking fringes
pattern emerges. Kinks and pinches
line up as matter's particles, the way
the threads in close-pressed layers of moiré
ripple into loops and dots — motion held
in place by woven crossings. Warp, weft, weld.

Light, even from our feeblest lanterns, takes
its being from a field that permeates
the cosmos. For photons, all of space contracts
to *here*, while *now*'s quicksilver passage lasts
forever. This most ordinary magic
is conjured from an all-pervading fabric.

Indra's net hangs above the peaks
of his holy mountain — the shining pleats
of a tent of stars draped above the world
where every knot is fastened with a pearl
and every separate jewel in the mesh reflects
every other gem at every vertex.

Mere myth, perhaps. But let us consider
Earth as one such gem. Cerulean mirror
gleaming from this corner of the dark
galactic lattice. Alone, we think. Apart.
And yet a world reflecting millions
of such worlds — limitless grid of brilliance
that, like the studded, crystalline arrays
of an insect's compound eye, surveys
the whole in parts. Indra's gift to us: to see,
in one small pearl, the gemmed immensity.

10.
   ... changed through all, and yet in all the same

Hope humbly then. At times it might be pleasant
to think of earth without ourselves — a present
where forest stretches, undisturbed, in place
of city. Where winding rivers trace
other paths than highway and the roar
of traffic's silenced. As it was before
(or might be after). An Edenic state
fenced by a giant *No Admissions* gate.

But we are here, and we cannot help it
that — even if we walked in shoes of velvet —
we disturb its threads with every footstep
and feel returning tremors from the web.

We are not atoms in an emptiness.
We're entangled, markings in a palimpsest

that's written over, time and time again,
by equations of the universal theorems
that underpin the cosmos and preserve
its balance — conservation laws that curve
matter into energy yet maintain
some essential quantity the same —
a measure that holds constant over time.

Within this fractal mesh of *equals* signs,
we are both large and small. Equivalent
in worth and intricacy to the ferment
of a galaxy's vast, spouting spirallings,
or a small, green, leaf-shaped frog. Things
do not get simpler just because of size.

(And while we're at it, it's time that we revised
that metaphor of 'cosmos as machine' —
the dated picture of a world that's made
of simple bits — the ultimate cliché,
complexity reduced to sophomoric.)

Instead, we have the combinatoric.
Potential connections in a single brain
build to numbers so much larger than
atoms in the universe; a single cell
beggars description with the pitch and swell
of its heaving interactome. A strand
of DNA within that cell expands
into a language able to encode
the tales of generations, a motherlode
of story.

                    Welcome, then, to this Great Chain
of Being, the net in which we claim
a place. If that place includes a soul
it's not likely to be individual —
a bodiless homunculus that floats around
without the laws of physics to impound

its mini-mind. No, it's far more likely
that soul is yet another force field, tightly
coupled to the world, a summed perspective.

We are time's derivative.
And for a little while, we are each a lens
in its compound eye. We might not unite
behind Pope's verse *Whatever is, is right.*
Still, whatever is, matters, in a wholeness where
everything is common and everything is rare.

The local globe

... We transform
the richly variegated niches
into starved soil for weedy species
like ourselves.

Windfall advisory

A whole TV channel to talk about the weather. Climate conditions twenty-four hours a day — we don't even need to look out the window. Did you see the weatherman point to his map today, to the spot down near Empress where he said there is a 'windfall advisory'? Imagine that, a windfall advisory! That's what you get living in Alberta. Oil gushing into your tank. Winning lottery tickets ripped from kiosks and swirling into the road. Dollar coins spinning along the streets, a clink of fallen leaves. They flatten the barley crop worse than a hailstorm, but you get the insurance payout on the spot. *Oh, no, I wouldn't want to live in Vancouver*, we say. *Six months of rain! I'll take a prairie windfall any day.*

There goes the neighbourhood

Magpie as neighbour. You've moved in,
hold your raucous parties, shout at the kids,
fix up your house — a slipshod, DIY
endeavour that always seems half-done. Twigs
strewn all around the yard.

We complain that you've forced out
the elegant kingbird couple
and that lovely warbler family who used
to ornament the neighbourhood,

        forgetting it wasn't you
who moved in first, altered the architecture
of poplar, hazelnut and reed-rimmed slough,
wild rose, stonecrop, berry bush.

        We ignore the fact
that you're the only ones prepared to cope with *us* —
to live off garbage bins out back,
plant your nests where predatory cars
go prowling by, and square up to the cat.

Guardians of Eden

Sky, the porcelain inside a Chinese bowl,
blue-glazed summer. A grey fence frames the bloom
of warm rose yarrow, lilies, baby's breath
innocent as Eden on this afternoon
with just enough breeze to tickle a tabby's ears.
A bee burrows in the lawn, backside
like a broad-beamed gardener in a black skirt
and gold plush sweater. The cats and I sit, satisfied,

until a grey intruder cat steps delicately
through the hedge and stops abruptly,
aware of three judgmental pairs of eyes.
'There's only so much room in Paradise,'
says my orange cat with twitching whiskers.
The stranger takes the point, and disappears.

Privacy acts

I've just been asked to sign a waiver
so the boarding-kennel manager
can hand out information to the vet
    about the cat
in an emergency — a recent edict of
    the privacy act.

I'm glad society's concerned about
protecting data on the cat's behalf.
He *is* a private animal, without a doubt,
    has never answered
questions as to where he's been, or where
    he got that bird.

    It's good to know the law protects
    one's right to be invisible
    beneath the bushes.

Our civic edifice is founded on the fact
that you can go inside your castle
with whatever bird you've caught
    and close the door.
The facts are no one's business but your own.
    No need to share.

Bird singularities

Bird mathematicians
struggle to calculate
those invisible walls
where the universe stops.

Space and time do loop-de-loops,
they trill sagaciously.
But there are singularities
where four dimensions of flight
intersect, contract to two,
and our equations are abruptly
banned from passing.

Passerines without classrooms
in which to acquire mathematics
become more practically aware.
These street-smart ones learn to shun
vertical planes that glimmer
with the lure of logic but
are based on false assumptions —
that air's equality
continues everywhere.

Dust to dust

A film of silver on the bookshelf,
fluffs of disaggregated substance
on the baseboards. Much of it is me —
      I'm shedding
seven million flakes of skin a minute,
a whole outer layer of myself
day by day. There's just so much dust
to dust. And it isn't only me.

      We're heading
gravewards faster than we think.
Seven billion people on the planet, plus
our dogs and cats, our cows, the pollen
from our mutated crops — and then
all that desiccated soil blown in
from desertified territories.
      A webbing
of soot and off-scourings, a scurf
that settles into every crevice, accretes
beyond the reach of rags and cleanser.
Dust to dust. The round world's winding sheet.

Annual grains

Agriculture's pornographic fact:
reproduction gets co-opted for consumption.
See that cornfield, tossing blonde tassels
and swelling into snug green corsets —
it's just the calculated wildness
of market forces. No truly wild plant spends
so much of its energy on sex,
on putting out, on hanging on
to seed heads that should shatter, scatter
small grains into earth's soft box.

And all this captive sexuality goes for what?
Youth, on its shallow roots, is taken
at the height of summer, loses all its profit
to pimped manipulation.

Demeter waits at the arrivals gate

with her arms full of flowers
and a welcome carpet that unscrolls
behind her to the parking lot
and the fields beyond that wait
for the tip of spring to be inserted
in their veins. She longs to clasp
her daughter to her brown heart
after all this time arranging visas
and permissions, releases from
the other powers. She needs to see
her girl, her red Persephone.

She did not expect
her daughter to return a pensioner —
too tired for revival of a world, riding up
the long dark escalator from below.
Her hands grey crepe,
gripping a suitcase packed with dried leaves
from old journals, a portmanteau
for which she fears that she has lost the key.

Red sky at . . .

January. Grey dawn sky.
The air is warm, unseasonable,

softening the snow that seemed invincible
just yesterday. The ravens *kronk*

in mild surprise, as if to thank
the god of thaw. The furnace stops

and in its wake of silence, thoughts
sift and stir, like cat hair

shifting in the quieted air.
Thoughts, of course, of gratitude

for ice's release and the beatitudes
fluted out by chickadees —

'Blessed are we
who have survived the minus-twenty

of the last harsh weeks.' But, gently,
the sky turns red — and that means 'warning.'

Not right now, not on this soft morning.
Danger is not so imminent

as that. But there are incidents
and auguries that show how change

is in the forecast. The winter's getting strange.
The future's birth-cord is being twisted

into being and we are complicit
in the spiral, the furnace starting up again
                    and I.

Climate change debate

Frozen. Solid. This slow, shallow stream
now paralyzed glass. The dog skitters
on its slick skin. Stubbed bulrushes lean
at fixed angles from its hard grasp.

Deep ribands cleave the clear mass
from surface to base. These silver fissures
are the borders of translucent territories.
On either side, molecules lock at odds,
orientations incompatible,
accidental. Allegiance chosen for them
by their neighbours.

It will require
spring's clearest signs of warming
to liquefy positions, offer mild rebuttal.

But then, too quickly, ice will pass
and the creek bed will be fired
to cracked, crazed mud, a fringe
of desiccated willow, gasping grass.

Badger

People come, they stay for a while,
they flourish, they build
and then they go.
It is their way.
— BADGER, *in* The Wind in the Willows

People come. Here there would have been
a slough, a marshy basin, liquid in spring,
sheen of snow melt, edges feathering
with reeds, Perhaps a heron, spearing
little fish — brassy minnows, finescale dace.
Certainly there would have been
poplar, aspen, chipping sparrow,
nesting passerines. *People come.*

I imagine voices. Small bands
passing to the river bend below,
returning year by year to campsites
along the secret pathways shared by deer.

They stay for a while. Newer comers
laid down rail lines here, and yards —
twenty-four tracks wide — for boxcars packed
close as cattle jostling in a pen.
Freight sheds, switching engines, and then
a stately station, column-fronted,
to welcome piled-up steamer trunks,
immigrants, battalions coming home
from theatres of war.

Soil laid waste. Clay stamped solid
by all that weight, littered
with rail ties soaked in creosote,
gravel, lumps of coal, the gleam
of oily puddles after rain.

Weeds pushed to the ragged edges —
sow thistle, toad flax, burdock.

They flourish, they build. Today
metal cranes swivel their long beaks
like enormous storks, fish up loads of iron
from the jagged marsh of girders
that rises round their waists.
Shunting shovels load dump trucks.
Huge piles of earth hauled off
to accommodate a bulging new arena,
towers and hotels where travellers
will open suitcases, shake out
temporary toothbrushes.

And then they go. I think of rust-belt cities,
their collapsing grandeurs —
theatres, cathedrals and the giant, empty
railway station where sun sifts
onto battered marble floors and arches
black with mould.

All the buried cities — Tall el-Hammam
Knossos, Abu Simbel, Troy.
And the unnamed middens, campsites, farmsteads.
Mounded over metres deep beneath
soil rebuilt from wind's continual freight
of dust and the patient disintegration of leaves.

I like to think of future roots
pushing through this paving,
of buckled towers becoming roosts for ravens,
the roof dome opening to the sky
like an ancient amphitheatre
and poplars standing, a chorus
of soft voices at centre stage.
A landscape of tunnels and hummocks
and badgers returning, wedge-shaped faces

peering out of burrows, earth moved
by their own strong claws.
    It is their way.

Mouse dreams

Cheese. Fat
kernels of grain. Smell
of old wood. Spider tickle.
Shred of dry grass in the nest.
Winter safety, when sky
is a low ceiling, a breathing
space between brown grass
and banked snow above.
Mouse pathways — quiet
arcades of pearled light
beneath frosted glass, an atrium
faintly warmed by fur and breath.
Tucked paws, a place
where we can sleep
with our bright eyes
closed.

Ratatoskr

Ratatoskr, squirrel, scurries
up and down the ash tree,
his world-axis, Yggdrasil,
heaven-wheel, winding spindle,
one tree that transects
the cosmos.

His scold-chatter carries gossip
and earthworm insults up
to the raven that alights, folds
wings like a wet umbrella,
black at the topmost branch.

This ash tree in my garden
grown big and bigger —
decades of girth-gain.
The crotch I once reached
on tiptoe, rescuing the cat,
now is far above my head.
The cat departed, a carton
of ashes among the worms,
wrapped by tree roots.

Earthworms not native here —
scrubbed from the landscape
ice ages ago. But now inching
back, rubber-bodied tubes,
agents of transmutation.

Ratatoskr scrambles, Raven
unfolds wings, the worms
chew through leaf-litter
and ashes. The ash tree
ever grows outward.

Waltz, wasp

Autumn wasp crawling
across the café window pane
       Sunlight's soft pollen
dusts the glass
            A waltz
starts to spill from speakers
      mounted on the wall
         above the menu
46 Dactylic pulse
    one two three
        melody
and it seems as though the wasp
      begins to dance
        swinging
on the downbeat
    turn again
      turn again
tracing arcs across the polished floor
    of a golden ballroom
      One two three
   harmony
imagined. It's an insect brain,
an arbitrary accidental music
    yet when the song and wasp
      lift off together
on a last high note
    I feel we are all
      caught in pattern,
   partners

A working world

And welcome to the billions muddling through ...

office hours

hickory dickory   click
of computer mice   from adjoining cubicles
      tick tick-tick tick

little chatter of mice teeth   behind
bland padded office dividers   three blind walls
and my back to the window
      tick-tick-tick

I long for the farmer's wife
to come down the hall with a carving knife
and chop computer cable
like the gristle of rodent tail
      tick-tick   tick-tick

and the clock strikes
one as going   so   damn   slow

I heard the bells ...

In the stomach-lurching morning after
the company's Christmas carouse,
the girl from down in accounting
wears earrings that are little bells
clipped to her earlobes with bows
of red-and-green enamel. 'I have such
a hangover,' she says and shakes her head
so the jewellery jingles. Co-workers,
wrung out with self-inflicted pain,
rouse themselves enough to wonder
'So why would you wear earrings
like bells?' Then put their heads back down
on grey metal desks that lurch to and fro,
relentless as the Great Bell of Bow,
and think there's no accounting
for what we do to ourselves.

Staff Christmas lunch

Celebration by decree, sign
here to bring desert, main dish,
or rolls and pickles.
Conscience puts you down
for spinach dip or salad plate,
though really you'd prefer
to bring the pop — a quick skate
through a convenience store
on your way in to work.

The conference table set
with plastic forks and margarine
in tubs. Combinations get
quirky. Chili and tuna casserole
swim together on your plate
There's always a surfeit
of the ordinary — cold
cuts, potato chips and bowls
of onion dip. And one
favourite dish that goes too soon
for everyone to get a share.

Even in the forced democracy
of holiday, we hone distinctions.
The manager assists with screwing lids
back on pickle jars, but
gets called off to the telephone.
It's the secretary who excels
at inventing creative decorations.
It's the receptionist who wears
the Santa hat and bells
and gets the job of throwing out
the mess. Poinsettia-printed napkins.
Plastic plates and cutlery.

The hierarchic choir re-ascends,
to higher floors for afternoon, and there is
an awful lot of waste at the end.

Free time

Here. Take it. No charge
for that hour or two after supper
tethered to the television
after work, after the dishes,
before bed and the alarm clock
ticking by your head.

Free for all. No accounts
need to be rendered
for this time when we sit
stunned by the news, never
straying too far. Like elephants
trained by tying their forelegs
to a tree. Eventually
intelligent beasts give up
on movements.

Free time. Political
broadcasts about productivity
in countries where they work
harder. And so to bed.
The clock ticking in your head.
Until unfree morning comes,
and tiny mahouts climb aboard
the massive, rolling shoulders.

Receptionist

This is what she controls.
The tapered shape, the tough colour
of lacquer.
           To keep them whole
she touches the world
only with the pads of her fingers,
tips placed carefully on keyboards
or telephone buttons.
The nails designed for
self-protection
themselves protected.

He watches her slit
envelopes. He controls
her time, chides her
for ten minutes' lateness
after lunch. Does he briefly
contemplate
the sharp red welts
her crimson claws could raise
on his middle-aged skin?
This kindly family man
with the plump, grey-flannel face.

She rasps away a tiny roughness
from the slight menace of acrylic.
Weapons never used
but non-retractable.

Bell curve

Bet you never thought we'd be this crazy
says the good-time-gal receptionist
at happy hour, when the room is hazy
with the blank statistics of getting pissed.
She's up for shooters upside-down, tequila
poured down her throat by a bored bartender —
she wants to set herself apart from the ordinary
day, to draw the line that represents a bender.

                    So here we go
crazy through the ordinary motions,
filling the mundane area below
the bell curve with our standard deviations.
            In vacuum flasks, there's a shortage of air.
            And it gets crowded in the bell jar.

The Gambler's Fallacy

Red Hot Jackpots — neon flames never stop
their scarlet jive above the VLT.
At the table where they're shooting craps
the white dice tumble. The croupier
rakes in the bets. The ring of oilfield workers
              flush with Friday's pay
(already flushed down this casino's drain)
pays to rattle the bones once more.
They think their luck will change

because they have been sucker-punched
by causality, this world where one thing leads
tightly to the next. But luck is no such
forged, linked chain. The dice toss heeds
nothing that went before — it starts from scratch —
cares nothing for the future hell to come
              when you go stumbling home.

After a morning spent in a visioning session with a well-paid consultant

In the lane in front of me, a tow truck
drags a little yellow school bus.
       Facing mournfully backwards
and caught on the intrusiveness of hook,
it is carted, ignominious,
and casts its headlamps downwards

at the pavement we are passing over.
I sympathise. Clearly, it desires
merely to be left alone,
to park its wheels wherever.
It doesn't want the hassle of repairs,
the conspicuous embarrassment of traffic cones.

And it simply doesn't want
to face the school kids or the schedules any longer,
or the constant need to be refuelled.

After a morning with the tow-truck consultant
(her reproaches levelled more in sorrow than in anger),
the air let out of tires, the pistons cooled,

I drive behind the bus in slow tandem
across the bridge — a rattling cortège
of mutual commiseration for the damage
inflicted, not by random
and unguided swipes of accident,
but by the hired guns of good intent.

Among the Magi

The waitress in the coffee shop
moves like a beautiful camel — that sway
on spindled thighs, the curving pout
that is a smile, the fringing lashes dropped
over her large, dark eyes.

It's the week past Christmas.
The decorations are tiring and the sun
is still far to the south.
He is fattening himself for migration,
the annual journey home.

The tinsel star is a tattered signal
in the slush-streaked window
to any wise men out there drifting
through the desert of snow
where some new thing lies a-birthing.

The waitress brings my coffee like a gift
from afar, as if she feels she knows
what I am looking for. 'No,'
I want to say as she sets down the cup.
'That's not it. But thank you anyway.
We'll have to keep on searching.'

Long division

... our plodding, conscious calculations
can barely cope with fluctuations
stirred into the paths of circling planets
by the smallest, palest, lunar fragments.

Catena
2.71828 1828 4590 4523 5360 2874 7135 2...

2   Crap shoot. Snake eyes. The double-twine
of chromosome, super-coiled, coils on coils

.   Point. Nit-pick of gene.

7   And then the downward plunge, the ravine
that splits a life away from fair beginnings.
Catenary curve — line of a rope bridge slung
above a chasm. Sickening sway
at its lowest point, suspended tension.
The chain of accident, curvilinear
that shapes a life.

1   Each life so wholly singular

8   inimitable outcome of equations —
universal patterns driving the particular.
Your brothers bend above you
in farewell. The summer twilight fades
to the murmured Kaddish. Your graven face
lays pale as feldspar. Three brothers, chains
of DNA — so sibling-similar
and yet unique. And the long chain
binding you to breath now broken.

2   Why do bad things happen
to biodiverse people? The random chosen?

8   Catenary — the dark arc mapping
the path of grief, its exponential plunge
from *no! This is not happening*
to the comprehended: *It is.*
The slowing increments of loss
when it can't get any worse, or
any better. The sad slog up, to stand
on something that approaches solid ground.

1   The curve by which decay and growth are bound

8   each to the other. The hand-me-down
    of inheritance. Three brothers,
    three sets of paired DNA.
    Two to the power of three — but, for you,
    a twist, untwist, unraveling skein
    that let the voices in. Distracting
    companionship, private and persistent,
    that stuck with you

2   splitting your brain's reality in two.
    The inner more convincing.

8   Some of your voices were kind
    and made us smile. Lawrence Welk
    inviting you to sing for him
    in Mendocino, California. And you sang
    for the nurses, *Welcome to my world.*
    But other voices were distressing —
    made disjunctive, angry claims
    on your attention. Strains sung

4   in a darker key. In response, you clung
    to a smattering of memories, like knots
    in the bridge's rope — boyhood's triumphs
    and hopeful trophies, the praise that made you proud.

5   Your prayer shawl is wrapped around your shroud.
    It is freshly ironed, and the knotted fringes
    lie smooth. Blue threads of memory that wind
    through a life. Your face now closed
    over the missing mystery of mind.

9   Prevalence of schizophrenia: nine
    in a thousand. Incidence of syndrome x,
    at age y — such calculations
    scatter data onto graphs. Discrete points
    freckling, filling out a range,
    shading in the shape of populations.

But who experiences life that way?
Each point merely plots a cross-section
of the line that runs from birth to end.

0 Ovum. Start again.

4 Chromosome four: a minuend
snipped shorter by subtraction of base pairs,
and a girl's body withers, decade
by decade, fibre by muscle fibre.

5 Now we place a spray of flowers beside her,
cerise and purple bougainvillea,
and a stuffed pink toy. Her breathing mask
can now be laid aside, and lilac silk
softens the harsh arc from scapula to hip.

2 Catenary. Traced by gravity's fingertip,
the curve of substance hanging by its own weight.

3 Curl, coil, turn. Fortune's wheel, Dame Fate,
Lady Luck. She does not toss the unlinked chance
of dice. She maps consequences of

5 initial conditions. The dystrophy,
invisible at first. Its continuance
a tightening cord that pulled her backbone
into a bow, winched her hips, locked
her into wheelchair. Disease confined

3 her body, but never touched her mind —
her glad passions, her love of purple,
the lines she drew and wrote.

6 How she loved Paris — the idea of it,
of flower markets with their notes
of violet and lavender, the style
of streets, romance of *arrondissments*
the curvature of Eiffel Tower, dressed
in the twinkling cloth of twilight.

0   The round nothing. Placeholder. Rest.

2   The awful intimacy chained
to the failing body, its excrements and mess.

8   The strain of lowering a coffin
into earth, labouring to restrain
its drop. How we hung on
so long to our end of the rope.
The dead weight of bad luck getting worse
until its ending.
The breaking ache. The tipping box
laden with what we have to love.

7   And the exhausted ones, the anchors standing
closest to the edge. Feet braced against
the pull. The compound growth of care,
its relentless decimal expansion.
Another journey to emergency.
Another jolt of hope, syringe of fear.
Another impossible decision:

4   Is it not time to hand Fate back her shears?
The blades we've tried so hard to keep from her,
binding their edges in a web
of surgical gauze, research, medicines.

7   And yet your brother murmurs in his dreams,
I never thought I'd see my brother's face
like that. I never thought I'd see him dead.
I never wanted him to die.
Catenary. It's the curve of memory,
her dreams of reaching Paris, their songs,
La vie en rose. Welcome to our world.

1   Each life a single digit that belongs

3   to the continuing. *Catena.*
From the Latin, 'chain'. The bonds
of stressed metal. The line necklaced, shining.

5   Catenary. Bow of acceptance, finding
its path across the gorge.
Spider silk in glinting wind, the swing
of phone lines suede with frost.
The messages we send. *Adieu. Adieu*

2   Two dead faces. A number that goes on
forever. Why you? Why you?

Zero divided by zero

There is no right answer.
The trains of logic crash, annihilate
certainty. Zero is just as good an answer
as one. Nothingness or loneliness.
There is no right answer.

The woman, Godelieve,
books the hour to annihilate
herself. An Antwerp euthanasia clinic.
A life too long, a brain too hurt.
Nothing is her answer.

The pastor sets himself on fire
in a Texas parking lot, annihilated
by grief for a world that cannot learn
kindness. He's tried everything else
finds no other answer.

'Mental illness' — that's one label
for the call to self-annihilation.
But is martyrdom, immolation,
never the sanest path, the kindest thing?
There is no right answer.

Zero divided by zero.
The past's black hole annihilated,
divided by the null of future.
Suicide's paradox: relief unfelt
by those who choose its answer.

Black holes are where
God divides by zero. Annihilated
light. All our singular arrangements
of matter reduced to one
undefinable answer.

My sister holds a vial
of danger, matter that could annihilate
her, but — divided into increments —
will let her sleep, let her breathe.
But there is no answer

for the anguish that strangles her,
for the losses that seem to annihilate
her past, divide her from a future
worth living. The vial's round mouth
seems an answer.

Saint Godelieve,
centuries ago you were annihilated,
strangled by the servants
of a cruel husband. Pray for us now
that we find an answer.

Your name means 'God's love.'
But God seems zero over zero — nil,
a void divided by a vacuum.
How could being, Oneness, ever
emerge as answer?

How can we live, here, Godelieve
in this universe where pain annihilates,
sometimes, all joy? The thought flares,
would it be easier to let her go?
Then the answer

No! My sister, oh my sister,
you terrify me. If you annihilate
this pain, then you divide us utterly.
You cannot be both zero and one.
There is no right answer.

Complex number plane

An envelope opens.
Your photograph falls
to the flat plane of the floor.
Your male face bisected
by slashing sunlight.

Lovers once. x-crossed.
So fraught a transit

through complexity,
so short a time. We fought
about how high to hang

the pictures, though our quarrels
were in truth, at root,
confused equations
of love and difference.
You were so different.

The men across the subway car
stared from me to you — our
clasped hands, your hair
crinkling to your shoulders,
your sockless ankles,

the giant hoops that hung
from your earlobes, round
as zeros, in those days when
one earring on a man
signalled a known code.

And one starer asked the other.
What do two earrings mean?

I did not mean you harm.
And yet I harmed. Then laughed
years later when I heard
you'd gone through knife and needle
to change your gender.

How did I miss that! I asked
remembering your dark
stubbled jaw, the thrust
of bulk, face and chest,
your obvious cock.

I missed so much. The sheer
ignorance of those dumb years
stuns me now. My young life
lived on such a narrow line —
just one dimension

where positive digits
ranged always to the right,
masculine and shining.
Their mirror yin —
the dark and feminine

negatives spreading
to the infinite left.
And at the centre
uncrossable zero,
definer of difference.

What do two earrings mean?

I had no idea then
that zero also doubles
as the round centre
of a second axis,
perpendicular extent

of numbers multiplied
by i — square root
of minus-one — a quantity
mathematicians termed
imaginary

through puzzled centuries.
'Impossible' but falling
out of ordinary
calculations, refusing
to disappear.

Proved newly useful
at last, constructing
a vertical spine
to open an expanded plane
of possibility.

What do two earrings mean?

Boys who want to wear
a Cinderella dress.
The hermaphrodite
folding a single body
around two genders.

Women whose upper lips
grow dark with hair.
Those inheriting
heterochromosomes
that don't line up.

I had no idea how
my limited arithmetic
clamped a lid
on complex planes.
We were stymied by our times,

by what could not be thought
in the world where we grew up.
(Your father's horror at the son
in a sparkly dress.) It was
a world of new techniques

deployed in old ways.
Surgical invention used
only to assign
confusing newborns
to the 'right' side

of a narrow number line.
Chemistry applied
to the straightening
of disorienting
orientations.

You stood at a verge
where wider understanding
could start to trail
discovery's slow curve.
But it's hard to be the first.

What could two earrings mean?

I wonder now — was
it really needed,
the surgery, the blasts
of antiandrogen, that
drastic subtraction?

Did you have to be
only woman? Could you not
have kept that cock,
reimagining yourself
as $1 + 1i$

and gone on loving
the women in yourself?
Did you just need to cut
a male face from your life —
that horrified father?

Or was it indeed
a horror of dysphoria
you had to escape?
entrapment in a body
that was not you,

and so you mapped yourself
as *minus 1+1i*
even as you went on
loving women
and never wanted men.

It saddens me to know
I'll never know
the answer, now buried
below history
and beyond reconstruction.

I feel a kind of shame. Not
that I failed to love you
— hearts map their own
complicated planes
and stubborn coordinates.

But I regret that laugh.
It did not rise from malice
— just a blurt
of surprise — yet it was
mindlessly unkind,

evidence of all
I did not see
while insisting pictures
must line up
at the level of my eye.

You could have been
an alternative axis
of vision. Forgive me
for living in my time,
accepting its blinkered limits.

Forgive me that I failed to see
your complex face.

Discounted annuals

... it's appropriate
for us to demonstrate humility.
We are not angels. But we're family.

Draft of a poem on 'inclusion'

There's a helium balloon up there.
A puffy silver star
pressing its cheek against the glass-peaked ceiling
over City Hall's grand stair.

It wants to be included in the larger air beyond.
That's what inclusion is — not letting in
but letting out.

We're tired of being kept inside
our glass-paned containers,
tired of hardness masquerading
as transparency.

Discounted annuals

The big-box grocery is an acreage of cars
crowded round
an immense interior — towering
aisles and piles
of cans. A factory stocked and re-stocked by
unnoticed hands.

I meet Larry at the back, near the deli counter
where we can buy
a slice of pizza for a dollar eighty-five
and bottled juice.
Then Larry leads me surely through the maze,
taking our lunch
away from the air-conditioned chill that's good
for groceries.

Hi, Larry, says the girl who shoves a trolley
wobbling high
with cereal boxes. And, *How's it going, Larry?*
asks the greeter
who has to check for evidence we've truly paid
for our pizza.
Great day for the race, the greeter jokes.
The human race,
he adds. He's told this one before.

We pass outside
beyond the awnings of a temporary garden —
discounted annuals
and sacks of perlite-lightened potting soil.
A picnic table
bakes in reflected heat from stucco walls
and softening asphalt
where we open up our sponge-foam boxes.

Yes, Larry is a little different from the rest
of us. And thus
his brain gets tagged with a discount sticker
by a world
that doesn't notice much beyond
the checkout counter.

For other shoppers, this giant store is just
a passageway
through a random aggregation of
have-a-nice-days.
But Larry knits it into a living village
where some folks
piss him off and many are his friends.

He stops to chat
with the pretty salesgirl in the garden shop.
I buy a pot
of frilled petunias, amazing in their ordinary
luminosity.
Larry carries my new flowers to the car for me

and it's a great day
for the human.

The hat

There is not another like it
in the whole known universe — and yet it's here
in this chain-franchised coffee shop, on the head
of a squat, lace-collared woman, her hair
fair, fine and fraying, greying. The hat
is an explosion of crochet —
woolly loops of rust and pumpkin
puffing upwards like the crust of a soufflé
or the pelt of an alien animal
from another galaxy. It conforms
to no known pattern. The bows
and bobbles with which it is adorned
will never be repeated in the wildly
multiplying combinatorics
of the world. The sullen young
snicker nearby, a tattooed chorus
of conventional rebellion with stapled lips
and earlobes. They'd never wear That Hat.
They wouldn't be caught dead.
They crave to set themselves apart,
unequivocally unique. But not
with that thing on their head.

The realms of asphodel

The girl concocting coffee in
        the Second Cup
has a face as pale as asphodel,
        skin perfect as petals,
fair gold hair caught up
        in a fist of elastic.
She is luminous as a flower
        beside a darkening path.

Nearby, waiting for her order
        of smoking cocoa,
a second young woman leans upon the edge
        of the counter, crowded
with cellophane-wrapped cookies,
        sweet bars, gift mugs —
duplicated goods pre-packaged
        for easy takeaway.

She watches the pretty barista
        with such longing
in her plain round face.
        To look like that?
To hold her in her arms? To brew
        a philtre that will call her
from caffeine's hissing machinery
        up to the lighted world?

The ghostly meadows of Elysium —
        the flowers that we pass
but may not take away.

Kind to a cat

The gurney's silver wheels go silent
as a stage illusionist's
machinery. The lobby of the seniors' building
chilled, empty. I stand,
an accidental honour guard
for the unmistakeable shape of stillness
taking leave.

She was kind to a cat.
Gave an old, ill-favoured beast a place
in her tiny apartment, filled
bowls of food and milk mornings and evenings.
One of the eccentric guild,
a solemn religiosity that made
her neighbours awkward.
            Still, she was kind.

I imagine the cat bewildered
by the prone form on the floor,
distressed, delicately nosing
the quilted dressing gown, curling
sorrowfully at her side.

But this is my willed illusion.
A cat knows better. Knows the utter gulf
when the familiar autonomy of cells
fails, when the fibrillation
of dissolving hearts goes still.

Instead, the cat goes to sleep
on the window sill. It knows
the shape pillowed on the floor
will be our kind
            no longer.

Child care

For a boy who cries too much —
Take dust from the hinges of the door,
knead it with his mother's milk,
place it on his head where the skull is soft.

Take the soft, wailing thing
shake it, shake it.
Close the door.

Take a child with measles to the crossroads.
Leave the disease behind.

Or leave a child. In a box
anonymously. Escape
its frail infection.

For a child subject to fits,
pull her through a bramble bush
where the evil spirit cannot follow

Or diagnose
FASD. Give her up
to goblins.

Old Anna
Children's clinic, Düsseldorf, 1915

On their rounds — doctors with their big round heads
in big round hats. Their vests are weighted down
by buttons and pocket watches. They stoop
over the foundling babies in their slatted beds —

the babies who fade like cut poppies in a jar
for all our cleaning, cleaning, cleaning
and shiny floors and bottles and boiled linen,
for all the smell of bleach and vinegar.

And this is Old Anna, says the Herr Doktor
Shlossman, in a hearty voice to justify
a fat old woman with spit all down her apron
hobbling along his shiny corridor.

*When we've done everything that science can
for an infant, we assign it to her care.
And always she succeeds.* Amazed, the foreign doctors
turn to look, as if I was a walking warming pan.

They have a fancy name for it, this fading away,
and go looking down their microscopes
for a little germ to cause it. As if you could see
a soul in a microscope. For this is what I say —

Old Anna's theory. A newborn's soul is light,
lighter than swansdown. Not sewn with buttons
or weighted with watches. When you cut the birth cord,
it will drift away — unless you stitch it tight

to that little soft body. A baby knows
when it has been cut loose. Decides to die.
It will survive on rocks or harsh sand, so long
as it has roots. Without them, nothing grows.

So I knit the soul in with my fat old sausage fingers —
tuck it into dimples, into their little palms
pink as clasped seashells, hold it firmly
against the tiny curved spine. And the child lingers,

reels in its floating, drifting soul like a kite
on a stronger string, buttons it down.
Looks up at Old Anna with its round eyes,
starts feeding with a nestling's hungry appetite.

But if I told Herr Doktor Shlossman this,
what would he say? *Now then, Anna,*
you may have something there. His pudding voice
that says he's humouring me, old superstitious

woman. What quaint ideas she talks.
He'd go on looking in his microscope
and sending learned letters to his foreigners,
while the babies go on drooping on their cut stalks

and their heads lie quiet on the white pillows,
like soft fallen petals on a quilt of snow.

The things we drag behind us

Oxygen tanks.
Small backpacks on wheels.
Reluctant children.
Dogs trying to take a poo.
All of history.
Humps, visible and invisible. And
poems, caught in the brain
and towed through our days
like a child's wooden train
made lovingly by hand.

Laundry hearts

... lessons in a murmuration
of starlings, shape-shifting veil of wings

This afternoon before the clocks turn back

Tonight we have an hour of the year
to live again. What hour will it be?

First snow sifts through the open gauze of air.
The street's softened prospect of what will be.

Weather forecasts differ in their prophesies.
How much snow tonight? How deep will it be?

In the grey light fading, colours pull in
around themselves, to concentrate on what will be.

Rowan berry, rose hip — their crimsons cluster
and pool on branches. They will be

warm lamps through winter. In these contracting days,
we need to know where lights will be.

What hour of the year's round clock
would I choose to live again tonight? It will be

this quiet one, when I do not know
        what will be.

In memoriam

It isn't easy
  to throw things in the sea
for the sea
  will bring them back again —
the ashes
  you scattered on the waves,
the flowers
  you flung as far as arms
would let you,
  bright, petalled parasols.
The sea returns
  such light detritus
with the clatter
  and recurrent roll of pebbles.
They come to rest
  just beyond your boots
among the scraps
  and bubbled straps of kelp
(translucent
  as the tinctured glass
of churches).
  Of course such things return. For who
could ever throw
  the weightless lace of memory
far enough away?

Battle River country

I am a scout in this country

> But the land's lay
> is silent as a ring of stones.

> I reconnoitre the next rise
> to find

> only the sigh of names
> driven away

What battles fought here?

> Skirmish of wind in the blades
> of the tall grass

> vegetation with its fierce names —
> *wolf willow, choke cherry*

> coyote's silver bugle
> lifted against the outpost sky

season of metal

a white line
trapped in black ore

grief tastes of metal
memories of rock

I hear the small sigh of rust
coming towards me on its soft feet

and a remote tremolo
as though a distant wire shivered

under tension

Laundry hearts

I'm held in place by such small pegs.
A little, specific dog. The man,
essential, who pours his cereal
each morning and leaves a spoonful
for the cat. The sister, whose face
I have never not known.

I feel I am a great wet sheet
flapping in the wind. Should
a single frail attachment fail,
should the coiled fulcrum at its heart
give way and snap apart,
then I might blow away entirely.

      My blowsy heart —
so narrowly fastened, when there are
so many laundry baskets scattered
with unused clothespins.
Put a few more on the line, I tell
myself. We could peg ourselves
more generously — become

a whole world criss-crossed
like the maze of laundry lines
between old tenements, up and up,
all the towels and pillowslips
billowing, catching for the breeze
that dries our cleansing, necessary tears.

Within, without

I want to read a winter's tale
of candlelight and nut-brown ale
and love returned.
Of weary roads and country inns
(the dark without, the light within)
where fires burn
in hissing grates with dropping coals.
A tale for sentimental souls
like you and me

who hope for hope and happy ends,
for simple deaths and fond amens.
Not tragedy.
Just sadness like a nut contained.
(The light without, the dark within
a walnut's shell.)

A tale of bitter bound by sweet.
The seed surrounded by its fruit.
The wooden bowl
filled with the autumn's apples. And
the grate's illumined ampersand
joining ash to flame.
I need a tale of winter's wind
(the dark without, the light within)
to read again.

In every tongue

Now coorie doon. The pull-out sofa pulled
from afternoon's hard leather
into night's cocoon, cuddled quilts and pillows.

In the grate, the fading coal's a loom
where crimson threads and stourie ash
are woven loosely into one another.

My mother's voice, the music of it.
The room illumined by firelight,
the towel-looped hot-water bottle
far down the bed, almost eluding
my foot, the click of clock, the droop
of all small creatures into sleep.

Coorie doon, my coo-me-doo.
The rounding of lips, of days. The furniture
of memory. A corner bureau
obscured in darkening wood.

Who says that any more? *Coorie doon.*
Who falls asleep in rooms perfumed
by coal fire, to a clock's ticked tune?

Coo-me-doo. Let us not lose
our douce endearments. Let them endure
in every tongue. *Oh, coorie doon.*

threshold

evening
and sky an inexplicable colour
like faith   or saints' eyes

you lean your face against the screen
a grey veil pressed into your skin
odour of dust and metal

its fine mesh too close
to focus   lines
blur like bars of a spectroscope
the pattern of something elemental

beyond is fragrance
                  and a bird on the fence
sits   so tiny   your eye fits it inside
one minute square

her song inhabits the garden

sun thread

solstice   sun stitched
to the high point of the sky   gold knot

chain stitch   daisy chain
small neat foot prints

lilac leads to lily   step by step
through iris   philadelphus   fat sweet peony

soul stitched into earth   by breath
the constant pouring forth of odour

feathered edges   grass's blanket stitch
lie down in its covering

running-stitch days   so long
so fast   sun's flashing needle

leaves woven with satin lines
and seed stitch scattered   everywhere

Foil

Today an arrow of edged joy.
A shaft. A passage from shadow into gold.

September's aureate branches toss — oro,
aurora, foil of leaves. Bright guild.

An aching blade slips below my ribs.
The sky's blue hook. The oak's black gall.

A topaz, faceted to trap light's glance
in its bevelled cell.

Clematis chimes its hours from yellow bells.
Samaras clatter on ash. Time is called.

Circadian Arcadias

Zeitgeber light. All our green clocks
begin to whirr. Molecular tick, protein tock.

Spring rises from within, endogenous
rhythmic shifts, a secret tic

released by light. Bare branches, blind,
see sun's change. The inner talk

of cryptochrome and melanopsin, rise and fall
of cycles, dawn to dusk, root to tip.

April's relentless pastorale. Caragana buds
ruff out, silver lambs. Sparrow cross-talk.

Even through the icy nights, ice wants to melt
from within. Water's constant drip and tick.

Et in Arcadia ego. Spring's alarm sounds brng...
brng... Bring your dead sticks and stalks.

Time will bury them in green.

The poet's handbook of cognitive illusions

All this with a brain
that doesn't realize its gaps and patches —
the leaps, elided details, makeshift matches
inherent in the maps it sketches.

Hallucinating the muse

Her name is Paracusis.

Scalp lathered, water plastering
my shoulders and the shower stall's hard tiles,
I pause — I think I've heard the phone.
Or its faint ghost. An old illusion.

I should know better than to fall
for it — it's too faint to be real,
a trick of porcelain and plastic curtain
that wrap me like a shell
held around an ear and murmuring,
falsely, of oceans.

Still I turn the water off
to listen, just in case the universe
is trying to reach me. But no.
I've missed the call.

pronominal

the cat does not say 'I'

he does not have to say
    I chase the toy mouse

oh to live like that
without the constant whirl
in my head   the daft-eyed noisemaker

to pounce, to feel
    I am pouncing

sure and un
  pronoun
     ceably

Pathetic fallacy

This late spring snow dump longs to melt
in the sunlight's strengthening arms. Cumulus
puffs loosen against the blue, move themselves
to illustrate the map of spring's
cerulean intentions. Streamlets seem
to follow us with newsy chatter
about the sky's high happenings.
     As if
the way the weather drifts around the world
showed motive. It's April's fool to us —
to speak of springtime as if it will express
attentiveness to our pathetic
need for company.
     As if.

Pareidolia

Giant yellow lady's slippers, orchids
of the north, lift their *Little Lulu* faces
in the shade. Paired, spiralled sepals
coil flirtatious ringlets as they tilt
their big chins. Beside
the tiny smiles of delicate forget-me-nots,
they are the ugly stepsisters —
les belles-soeurs, les belles laides.

But now that my neuronal nets have fused
soft-pouched flower with a face, I'm stuck
with that cartoon, cannot undo its click.
Our eyes — such exigent interpreters
of world — insist on faces. In clouds, hills, stars —
our necessary, narcissistic metaphors.

The Texas sharpshooter fallacy

Blam! The poet blasts a page
with words like buckshot spraying
the shingled side of a barn.
Of course the shot lands somewhere —
the barn's too big to miss entirely.

'That's what I meant to hit,'
the poet says. 'Yep, that's
what I was aiming at.'

She draws the circle of a title
around the accidental scatter.

Readers study the marks for pattern.
Is it meaningful? Does it say
anything coherent on the subject of
locked doors or hay or splintered boards?

'Must be something in it,'
they murmur, scratching at their heads.
The poet re-loads.

Necker cube illusion

perceptual toggle, perpetual
       flop-flips of cortex
trying to resolve contradiction —
       back-and-forth saccades
between 'see this,'
       'see that'

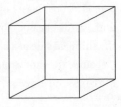

true/false trip-up, simultaneous
       and binary

what does this construction mean?
does it prove x
       or hidden *Why?*

       does that corner
tilt towards the sky
       or gutter?

(does what the poet's doing
       even matter?
yes and/or no)

and yet we have to hold
       ambiguity together, one
assemblage of lines
       that points to everything, all
at once.

Confabulation

A perfect Judge will read each Work of Wit
With the same Spirit that its Author writ.
— ALEXANDER POPE, 'An Essay on Criticism'

The sneer embattled — that lifting shrug
of *levator labii*, dismissive tug
of facial muscle that accompanies
argumentative disdain. Litanies
of poetic fault and failing then pour forth
severe decrees — for, of course
all muscles hook together. This specific one
is closely linked to those that lift a pen.

If such-and-such a bard is over-valued
on literature's year-end statement — well, you'd
better ring the bell, expose the fraud,
call in the poetry accounting squad.
It isn't personal. It's simply clear
that sound accounting standards here
are being violated. Expert notes
can point out where a reputation floats
on puffery and nepotistic plaudits.
We're simply saying there should be an audit.

Of course it's personal. To criticize
is human. We may insist our views derive
from logic, unsullied by emotion,
rational as Newton's Laws of Motion.
But that's the essence of confabulation —
the brain's compulsion just *to make things up.*

The cortex runs amok
with narrative. It isn't only victims
of a stroke, damaged brains insisting
their limbs are fine, could move, are only resting.

It's all of us and every day, investing
so much neural circuitry to render
our lives coherent in our mental blenders.

And of course we feel that itching to deride.
It's called 'gossip.' No human tribe survived
without poetry and scuttlebutt, those arts
of language. The social order's Cuisinart
cuts us down to size with blades of blather.
It's our advance on banging up each other
with antlers, or baring pointed canines
(the sneer's precursor). The grapevine's
just as much an instrument of status
as a thumped chest.

            So we target-practice
on some other poet's opus,
but then confabulate our motives
for bashing their stanzas silly.
            Face it. We aren't really
offering correction or a tonic
for moribund poetics and the chronic
fatigue syndromes of verse. When we critique
soporific lyric, the pedant-speak
of elitist bafflegab, hand-me-down cliché,
the drivel of confessional display
or post-avant anything, it's just because
anything that I don't like is therefore flawed
by definition.

            We'll never overcome
The lure of literary gossip — it's too much fun.
And necessary. It's how we pay attention.
Nor should we propose complete suspension
of critical thought, kindness grinding up
informed opinion into infant pap
the bland consistency of mushy peas.
            That's not my plea.

But we could put away our chopping blades,
the whirling Cuisinartry of crusades.
Let's throttle back our primate egos,
and their glib confabulation: *My torpedoes*
are simply launched in poetry's defence.

Close the spreadsheets, the pretence
we're able to enumerate and rank
quite rationally, can tabulate the bank
balance owing. Instead, let's go for balance

by looking at a poem whole, in silence
first. To see its shape, not maimed
by spite or bias, nor as a war game's
target, a tiny blob that we could blast
clear out of literature's wide range. *So vast*
is Art, so narrow human wit.
        Dear scribes, let us subdue that lifting lip.

The League of Poets Burial Society

... or maybe the poets will bury me,
or the librarians
— *RONNA BLOOM*

Membership in the League entitles you
to three linear inches of bookshelf space,
an urn engraved with text (from your best review)
to hold the ashes of remaindered pages,

and funeral obsequies held between the shelves
of a used-book store, the mourners sure
they could have penned better eulogies themselves
than you did — while they sip *vin ordinaire*

and nibble cubes of cheese like frugal mice
in literature's littlest church. Membership
in the Poets' Burial Society comes at a modest price.
Integuments include a nice jacket of dust.

A bookplate with your title will remain
ad aeternitatis, on a crowded shelf
below a window gloried by the rosy stain
of glass that's inlaid with the name of someone else.

Epilogue

... changed through all, and yet in all the same

Cledonism

What will be? I ask the god,
her breath restless in the branches.
but do not wait for her green answers.

Instead I clap my hands to my ears
and hurry to the human marketplace —
to my doctor, my investment broker,

to the checkout clerk or to the mailman
walking up the sidewalk with his bag
of envelopes and oracles.

I seek the single word that makes all clear,
the word that will inscribe a circle
around the possibilities of future.

A word like *fame* or *comfort.*
A word like *cure, secure, release.*
A word like *capital* or *teeth.*

Back in her June temple, the god's rejoinders
are tangential, daisy-chained,
interleaved with rustle. She murmurs *change ...*

*chant and vagabondage. Astringent journeys ... change
contingent on pigeons ... enjambed
poundage of engines ... the endangered.*

Age, she whispers. *Pasturage and larch ...
Languages ... chickadee jargon ...
the scutcheon's argent margin ... chance*

*coinages engendered ... genome ...
... angels ... branching channel ...
Change.*

Notes

In medias res
This Latin phrase, meaning "in the middle of things," is used for a
drama that begins in the midst of action.

Welcome to the Anthropocene
This sequence is a response to Alexander Pope's 10-part "An Essay
on Man." The First Epistle of that work is imbued with 18th-century
science. The latest discoveries through the microscope and telescope
confirmed Pope's concept of the Great Chain of Being ascending
from the "microscopic eye" of flies to the supreme being who
ordains the Newtonian orbits of planets. For Pope, our big sin is
intellectual pride.

The "black-6 mouse" is the most widely used breed of laboratory
mouse. "Bottle 38" refers to the first discovery of the *dunce* line of
mutant fruit flies, which seem normal in all respects except for
being incapable of learning to associate an odour with a negative
experience like a shock.

Bird singularities
In mathematics, a singularity is a point where a mathematical
relationship (like a function or an equation) breaks down and
changes its nature. They are extreme points where we can't define
how a function will perform or a surface will extend.

Annual grains
Grains form the basis of the global food system, but the crops
grown now bear little resemblance to the perennial grasses that
were their wild ancestors. One of the major changes that came
about as a result of domestication is the development of "non-
shattering" seed heads; seeds stayed on the plant longer, reducing
the amount of grain lost before harvesting.

Ratatoskr
In Norse mythology, Ratatoskr is a squirrel who runs up and down
the great ash tree Yggdrasil at the centre of the world. He carries

messages between the eagle perched at the top and the "wyrm" that dwells beneath one of the tree's three roots.

The Gambler's Fallacy

This is the mistaken belief that, in a string of truly random and unrelated events (like a series of coin tosses), the next result depends on the previous ones. It leads to the conviction that if there has been an unusually high frequency of "heads" then the next toss must balance out that trend.

Catena

The subtitle represents the first 37 digits of the decimal expansion of e, the number which is the base of natural logarithms. It is also essential to the formula that draws the catenary curve, the family of curves followed by a cable or string suspended between two fixed points. (The catenary also incorporates the equation for exponential decay.)

"Catenary" derives from the Latin, *catena*, chain.

Complex number plane

i is the mathematical symbol for the square root of minus 1, a quantity that mathematicians considered "imaginary" for centuries, as there is no number that can be multiplied by itself to obtain -1.

The realms of asphodel

Asphodel is a handsome flowering plant with a rhizome that is edible but not very nourishing, and so was used only by the poor as food. This fact may have contributed to its legendary association with the underworld in ancient Greece.

Old Anna

The Children's clinic in Düsseldorf, (as in other foundling homes and institutions in the early 20th century) experienced high mortality as a result of what was then called "marasmus" — a condition in which babies with no diagnosable illnesses failed to thrive. Nurses at the Düsseldorf institution would give the worst cases to "Old Anna," a cleaning woman, who was essentially the cure

of last resort. Invariably the children given to her did survive. Dr.
Fritz Talbot of Boston visited the home and saw Anna with a baby
on her hip as she did her chores, and came to realize the missing
ingredient in care was touch.

Circadian Arcadias

Circadian rhythms are biological processes that oscillate over the
period of a day, and are found in living things throughout the
biosphere — plant, animal, microbe. These cycles are generated
from processes within the organism. However, they can be
adjusted by a *zeitgeber* (German, "time-giver"), an external cue like
light or temperature that connects the inner rhythm to the local
environment.

Et in Arcadia ego translates as "even in Arcadia, there am I" and is
a *memento mori*, a reminder of mortality.

Hallucinating the muse

Paracusis is an auditory illusion or hallucination.

Pronominal

The *Oxford English Dictionary*'s definition of pronominal is "serving
to indicate things instead of naming them."

Pareidolia

Pareidolia is the psychological tendency to interpret a vague or
random image as something significant — most frequently, to see
a human face.

The Texas sharpshooter fallacy

A fallacy in interpreting data that arises when someone has a large
volume of data but focuses only on a subset of it.

Confabulation

A condition in which individuals with brain damage tell stories
and provide explanations without realizing they are not true.
The *levator labii superioris alaeque nasi* is a muscle from the nose
to the upper lip that enables the sneer or snarl. (*Levator labii* is
pronounced leh-VAT-or LA-bee-EYE.)

Cledonism

An ancient method of divination in which a petitioner listened
for the god's answer to her question among the chance words of
pedestrians.

Acknowledgements

The author thanks the editors of the following publications in which some of these poems have appeared (sometimes in slightly different versions).

Literary magazines

The Dalhousie Review: "Annual grains"
Eighteen Bridges: "The Hat"
FreeFall: "The realms of asphodel"; "Bell curve"
Kenyon Review: "Welcome to the Anthropocene (Sections 1—4)"
The New Quarterly: "In memoriam"
PRISM *International*: "This afternoon before the clocks turn back"
Vallum: Contemporary Poetry: "Privacy acts"

E-zines

Cede: "Circadian Arcadias"; "Waltz, wasp"
London Grip: "Dust to dust"
Truck: "Foil"
The Trumpeter: "Sun thread"

"There goes the neighbourhood" was selected for the Academy of American Poets' 2016 "Poem in your Pocket Day" collection.

"Foil" became one of the poems used for the Fibre Arts Network's *Ekphrastic* project, with exhibits across Canada and in the United States.

"Discounted annuals" is for Larry Marcotte.

Thanks also to the wonderful team at the University of Alberta Press, especially editor Peter Midgley. And thanks to the poets who have helped with suggestions on individual poems, including members of my long-time poetry group (Ruth Anderson Donovan, Gary Garrison, Ellen Kartz, Deborah Lawson, Naomi McIlwraith, Ella Zeltserman) and Luke Daly. Exchanging ideas with other poets is a great privilege and pleasure.

Other Titles from The University of Alberta Press

Standard candles

ALICE MAJOR

Scottish-Canadian poet links cosmology, mythology, and the human heart in a range of poetic forms.

Robert Kroetsch Series

Intersecting Sets

A Poet Looks at Science

ALICE MAJOR

Part memoir, part ars poetica, Alice Major discusses science with characteristic gleaming perspicacity.

The Office Tower Tales

ALICE MAJOR

Alice Major exemplifies the redemptive power of story in this ambitiously allusive long poem.

cuRRents Series

More information at www.uap.ualberta.ca